Phloem Transport in Plants

A Series of Books in Biology

Editors: Donald Kennedy and Roderic B. Park

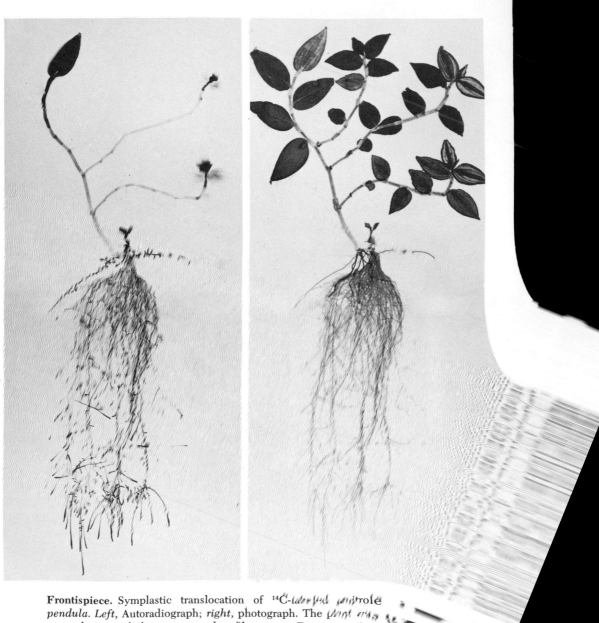

Frontispiece. Symplastic translocation of ^{14}C-labeled amitrole
pendula. *Left*, Autoradiograph; *right*, photograph. The plant was
ture solution with the roots spread on filter paper. Dosage was 0/2
time was 4 days.